Look-Alike Animals

IS IT A DOLPHIN OR A PORPOISE?

by Anita Nahta Amin

PEBBLE
a capstone imprint

Pebble Sprout is published by Pebble, an imprint of Capstone.
1710 Roe Crest Drive
North Mankato, Minnesota 56003
www.capstonepub.com

Library of Congress Cataloging-in-Publication Data
Names: Amin, Anita Nahta, author.
Title: Is it a dolphin or a porpoise? / by Anita Nahta Amin.
Description: North Mankato, Minnesota : Pebble, [2022] | Series: Look-alike animals | Audience: Ages 5-8 | Audience: Grades 2-3 | Summary: "You see a fin poke out of the water. It's a big gray animal with a friendly face. It looks like a dolphin, but is it a porpoise? Dolphins and porpoises are similar, but they have some important differences. Find out how their fins, mouths, teeth, and behaviors can all help you tell these two animal look-alikes apart. Filled with stunning photos and playful text, early learners will be delighted as each page turns"-- Provided by publisher.
Identifiers: LCCN 2021004139 (print) | LCCN 2021004140 (ebook) | ISBN 9781663908551 (hardcover) | ISBN 9781663908520 (pdf) | ISBN 9781663908544 (kindle edition)
Subjects: LCSH: Dolphins--Juvenile literature. | Porpoises--Juvenile literature.
Classification: LCC QL737.C432 A54 2022 (print) | LCC QL737.C432 (ebook) DDC 599.53--dc23

Image Credits
Alamy: Nature Picture Library, 17, Solvin Zankl, 19, 24–25, WaterFrame, 26, Wildlife GmbH, 20–21; iStockphoto: BrendanHunter, cover (bottom), 11, 12, Gerald Corsi, 5, 14–15 (top); Newscom: blickwinkel/picture alliance/F. Hecker, 9, imageBROKER/Willi Rolfes, 7; Shutterstock: Alexey Sokolov, 13, Anna_Kova (design element), cover (middle) and throughout, bearacreative, 18, Brandon Bourdages, 10, Coulanges, 4, Elena Larina, 23, Four Oaks, 16, frank ungrad, 30, grafxart, 28–29, michaelgeyer photography, 6, Michelle de Villiers, cover (top), Mike Price, 27, Miles Away Photography, 14–15 (bottom), Sergey Uryadnikov, 8, vkilikov, 22–23, WismarPhoto, 3, yeshaya dinerstein, 31

Editorial Credits
Editor: Carrie Sheely; Designer: Elyse White; Media Researcher: Svetlana Zhurkin; Production Specialist: Laura Manthe

Printed in the United States   4723

A fin shaped like a triangle juts out of the sea. "A **dolphin!**" you shout. But the animal doesn't stay. It swims away. "It's a **porpoise,**" someone says.

Dolphins and **porpoises** look a lot alike. Both are large, dark-colored animals that live in water. But they're different too. Let's compare these smooth swimmers!

pink river dolphin

There are **40 types** of dolphins and **seven types** of porpoises. Each kind has its own living area, or range.

Where do they live?

Some, such as Dall's porpoises, live only in the **ocean.** Others, like pink river dolphins, live only in **freshwater rivers** and **lakes.**

Dall's porpoise

dolphin

Something seems
fishy!

But it's not the porpoise's fault.

Dolphins can't be blamed either.

Neither are **fish!**

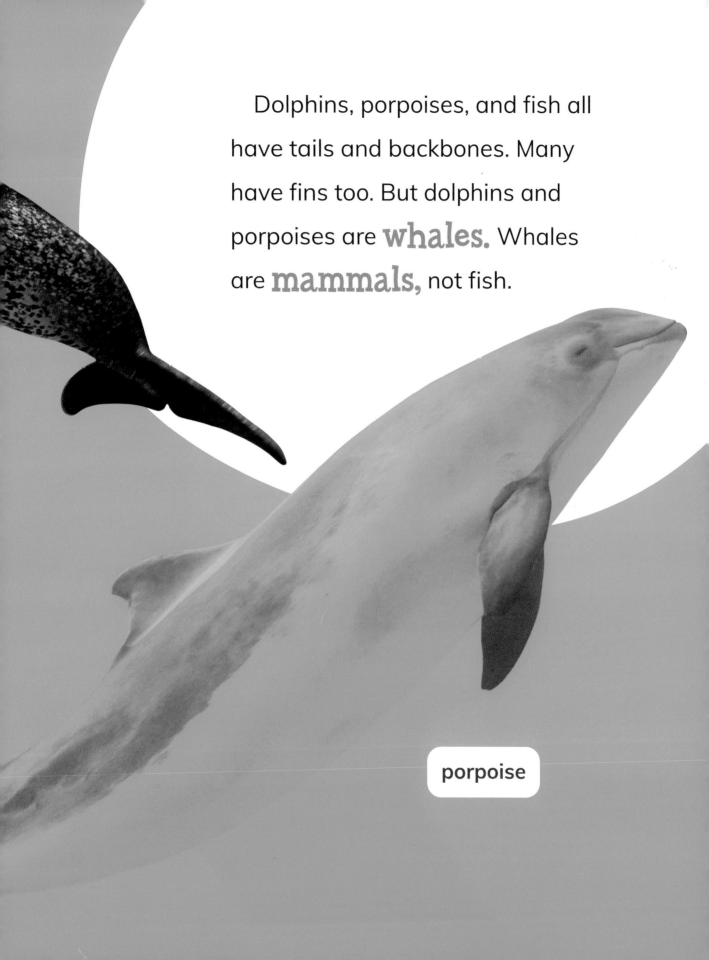

Dolphins, porpoises, and fish all have tails and backbones. Many have fins too. But dolphins and porpoises are **whales.** Whales are **mammals,** not fish.

porpoise

See the **fin** sticking up on this animal's **back?** It is the

dorsal fin.

dolphin dorsal fin

porpoise dorsal fin

Most porpoises have dorsal fins shaped like **triangles.** Most dolphins have **curved** dorsal fins. But this isn't always true. Some dolphins and porpoises **don't have any** dorsal fins!

dolphin

Open wide!

Most dolphins have **long jaws** that stick out. This part is called a **rostrum.** Porpoises have **small mouths** that don't stick out.

Dolphin **teeth** are **pointy.** They look like upside-down cones. Porpoise **teeth** are **rounded** and **flat.** They look like upside-down pears.

porpoise

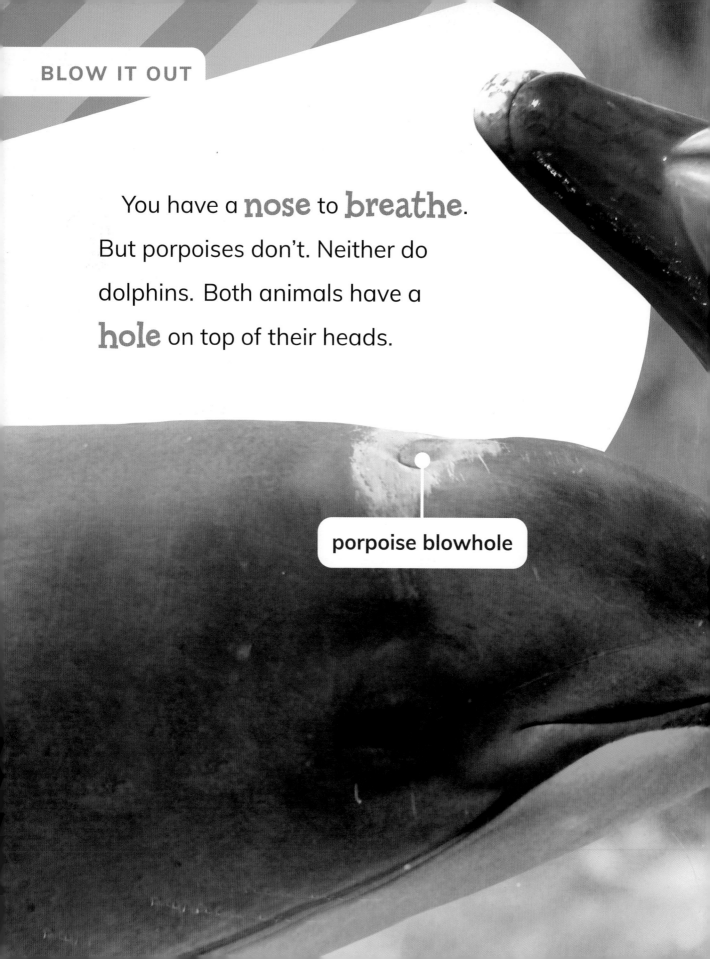

You have a **nose** to **breathe**. But porpoises don't. Neither do dolphins. Both animals have a **hole** on top of their heads.

porpoise blowhole

dolphin blowhole

They breathe with their blowhole.

They keep it closed underwater.

They open it when they come up for air.

Who is
bigger?

Dolphins would win this contest!

Most dolphins are less than

10 feet (3 meters) long. But the

largest dolphin, the orca, can be

32 feet (9.8 m) long!

Dall's porpoise

The largest **porpoises** are **Dall's porpoises.** They are about **8 feet (2.5 m)** long. Although they're smaller, porpoises are usually **wider** than dolphins.

orca

dolphin

Leap!
Splash!
Repeat!

Dolphins can **leap high** into the air. They can **flip and twist** too. Porpoises stay in the water and rarely leap.

porpoise

dolphins

Dolphins love to play with friends.

Their group, or **pod**, can have more than **1,000** dolphins! Dolphins are more **social** toward people too. They sometimes get close to people in the wild.

Porpoises are less social than dolphins.

They swim **alone** or in **small groups.** They also are **shy** and usually stay away from people.

porpoise

Neither animal can see far.

Do they need glasses? No, they need their **melon.** This organ sits inside their **foreheads.**

melon

porpoise

The **melon** sends out **sound beams.** The sound **echoes** back from objects. The echoes tell the animals where the objects are located. This helps dolphins and porpoises **hunt** and **find** their way around.

dolphins

Click, click!

Dolphins click, squeal, whistle, and more. This is how they **talk.** For example, a mother's buzzing tells a baby to stay close. Dolphins also use **body signals** to **communicate.** They might slap their tails to warn others of danger.

Porpoises mostly **click.** Different clicks have different meanings. An **angry** porpoise might repeat **fast clicks.**

dolphins

That **porpoise** seems quiet.

But maybe it's **talking.** You just can't hear it. You need a special machine to hear it.

Human ears hear only certain sounds. You **can hear** most dolphin sounds. But you **can't hear** most porpoise sounds.

porpoise

Dolphins and porpoises are smart.

People often **train** dolphins at aquariums to do **tricks.** Porpoises are **harder** to **train.** They are **shy.**

porpoise

dolphins

Dolphins show off their **brainpower** by using tools. Can you catch fish with a shell? Some dolphins can. They **teach** other dolphins too.

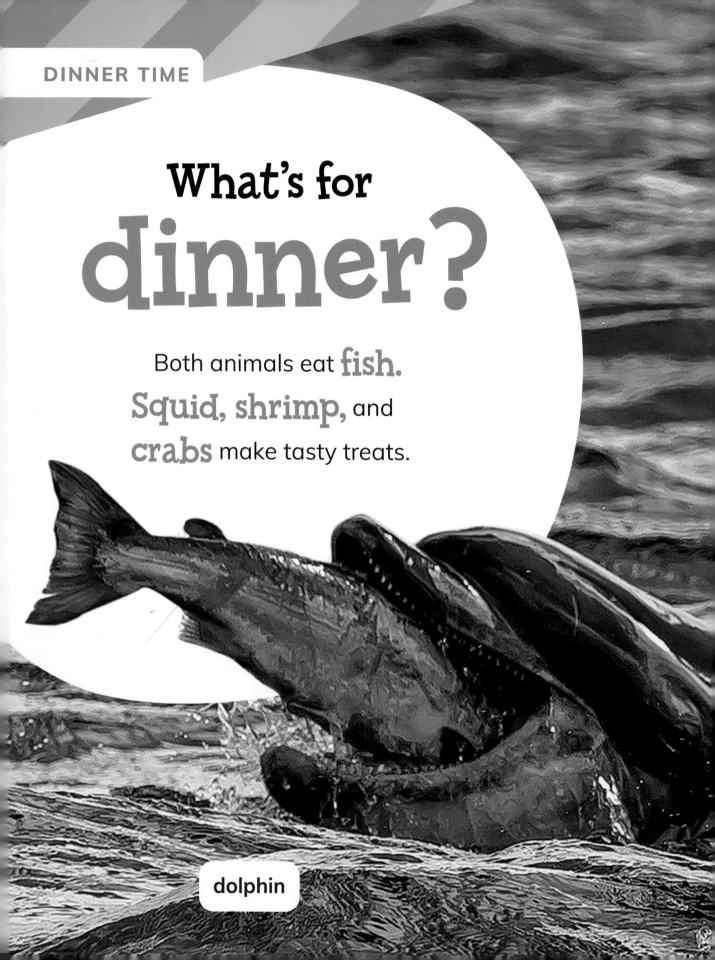

What's for dinner?

Both animals eat **fish**. **Squid, shrimp,** and **crabs** make tasty treats.

dolphin

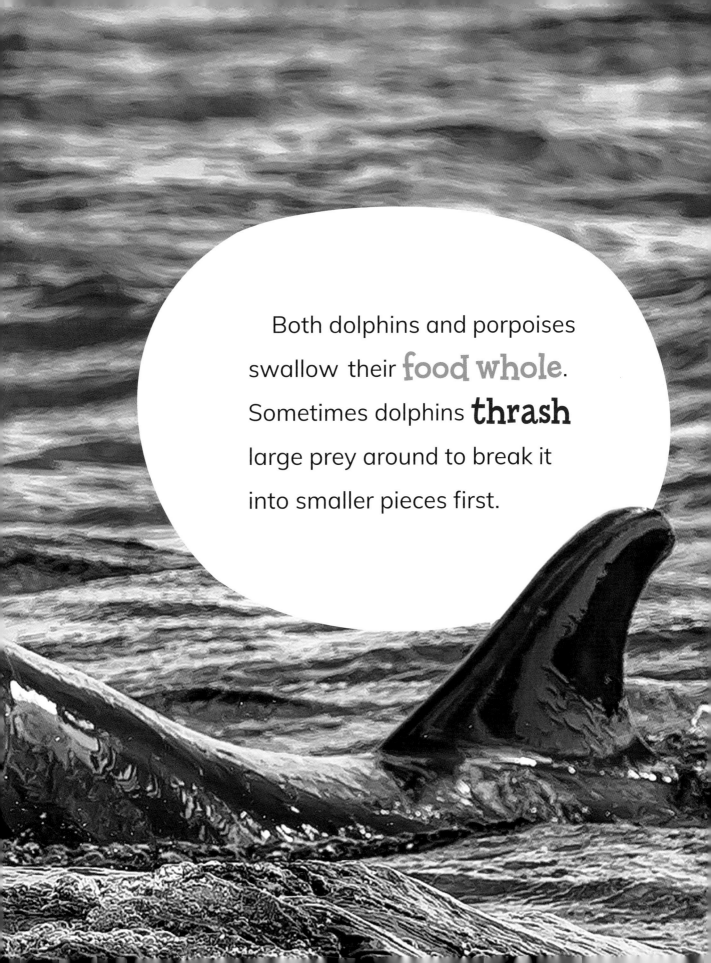

Both dolphins and porpoises swallow their food whole. Sometimes dolphins thrash large prey around to break it into smaller pieces first.

Dolphins and porpoises sleep with one eye open!

They stay **half awake** so they can watch for **danger.** They also stay partly awake to **breathe.** You don't tell yourself to breathe. But they have to. They don't breathe automatically.

IS IT A DOLPHIN OR A PORPOISE?

1. A triangular fin juts out of the water. Is it a dolphin or a porpoise?

2. It leaps out of the water. It twists in the air. Then splash! Is it a dolphin or a porpoise?

3. It has long jaws that stick out from its head. Is it a dolphin or a porpoise?

4. It holds a shell to catch fish to eat. Is it a dolphin or a porpoise?

5. It mostly clicks. Is it a dolphin or a porpoise?

Answer Key:
1. porpoise 2. dolphin 3. dolphin 4. dolphin 5. porpoise